THE SHANTY BOOK PART II

Richard Runciman Terry

A Classic Collection of Sailor Sea Songs, Chanteys, and Work Tunes

Lyric Legacy Historic Edition

Lyric Legacy

Originally published in 1926 by Richard Runciman Terry.

Lyric Legacy Historic Edition ISBN
978-1-64389-183-5 (pbk)

The Shanty Book
Part II
Sailor Shanties

(Curwen Edition 2952)

Collected and Edited with Pianoforte Accompaniment, by SIR RICHARD RUNCIMAN TERRY, with a Foreword by SIR WALTER RUNCIMAN, Bart.

London: J. Curwen & Sons Ltd., 24 Berners Street, W.1
Philadelphia: Curwen Inc., 1701 Chestnut Street

Vocal Edition, containing melody only (C.E. 6340), price 1s. 6d.
The Shanty Book, Part I, Curwen Edition 295
Made in England

TO MY SON, PATRICK HUGH, WHO HAS ALREADY
ELECTED TO REVIVE THE FAMILY
TRADITION AND FOLLOW
THE SEA

*FOREWORD

By SIR WALTER RUNCIMAN

IT is sometimes difficult for old sailors like myself to realize that these fine shanty tunes—so fascinating to the musician, and which no sailor can hear without emotion—died out with the sailing vessel, and now belong to a chapter of maritime history that is definitely closed. They will never more be heard on the face of the waters, but it is well that they should be preserved with reverent care, as befits a legacy from the generation of seamen that came to an end with the stately vessels they manned with such skill and resource.

In speech, the old-time 'shellback' was notoriously reticent—almost inarticulate; but in song he found self-expression, and all the romance and poetry of the sea are breathed into his shanties, where simple childlike sentimentality alternates with the Rabelaisian humour of the grown man. Whatever landsmen may think about shanty words—with their cheerful inconsequence, or light-hearted coarseness—there can be no two opinions about the tunes, which, as folk-music, are a national asset.

I know, of course, that several shanty collections are in the market, but as a sailor I am bound to say that only one—Capt. W. B. Whall's 'Sea Songs, Ships, and Shanties'—can be regarded as authoritative. Only a portion of Capt. Whall's delightful book is devoted to shanties, of which he prints the melodies only (without accompaniment); and of these he does not profess to give more than those he himself learnt at sea. I am glad, therefore, to welcome Messrs. Curwen's project of a wide and representative collection. Dr. Terry's qualifications as editor are exceptional, since he was reared in an environment of nineteenth-century seamen, and is the only landsman I have met who is able to render shanties as the old seamen did. I am not musician enough to criticize his pianoforte accompaniments, but I can vouch for the authenticity of the *melodies* as he presents them, untampered with in any way.

WALTER RUNCIMAN.

Shoreston Hall,
Chathill, 1921.

* Reprinted from 'The Shanty Book' Part I

CONTENTS

ALPHABETICAL INDEX OF SHANTIES

INTRODUCTION

As I am shortly publishing a historical and critical study of sea shanties, there is no necessity for this introduction to be other than brief.

During the five years which have elapsed since the publication of Part I there has been something of a boom in shanties. Old collections have been refurbished and put on the market again; new ones have appeared both in England and America. But when all has been said and done, Capt. W. B. Whall's *Sea Songs, Ships, and Shanties* still maintains its pre-eminence as the one authoritative book on the subject—the one and only book in which a sailor will find nothing erroneous or even disputable.

With the exception of Joanna C. Colcord's *Roll and Go* (published last year), the American collections are compilations from other people's work. But Miss Colcord speaks with the authority of one who was born at sea in the cabin of a sailing ship of which her father was captain. She claims descent from five generations of deep-water seamen. She spent the first eighteen years of her life in her father's ship, sailing with him on China voyages, 'knowing none but seamen, seeing nothing but ships, and ports and oceans.' Her book will prove even more useful when it is purged of the inaccuracies which seem to indicate hasty or careless preparation for the Press. I note two examples since they concern myself (they are typical of the rest).

On page 23, she says:

> Perhaps some idea of the difficulties in the way of the collector of shanties may be gained from the fact that this same shanty appears in the Tozer collection under the name 'The Chanty-Man's Song', the first line being 'I'm chanty-man of the working party', in Bullen's as 'Oh, what did you give for your fine leg of mutton?' and in Terry's as 'The Wild Goose Shanty', bringing in the mysterious 'Wild Goose Nation' which recurs in several British shanties. All of these versions use practically the same chorus; but none makes mention of the quest for huckleberries.

The answer to this is that not only do Whall, Bullen, Sharp, and myself include the 'huckleberry' verse, but in my 'Notes on the Shanties' in Part I of this collection, I wrote (page xiv) 'the verse about huckleberry hunting was rarely omitted.'

Again (on page 9)—speaking of 'The Black Ball Line' shanty—she says:

> As happened to so many of the old shanties, this one was modernized in later years; and the version given by Terry works a ship from Liverpool to Mobile and brings her back loaded with cotton—a voyage no Blackballer ever made.

The answer to this is that *my* version of 'The Black Ball Line' is now printed *for the first time in this volume*. By what species of clairvoyance Miss Colcord could scent inaccuracy in a version nearly two years before it was published beats me entirely. As a matter of fact, the version which makes a Blackballer perform so remarkable a voyage will be found on page 26 of Sharp's *English Folk Chanteys*.

Of the English collections published during the past five years, the one which is not a compilation from printed sources is *Six Sea Shanties*, by A. Whitehead and Taylor Harris, published by Messrs. Boosey and Co., in which the authors have taken their tunes down from the actual singing of a seaman.

One feature of this volume (which was absent from Part I) is the capstan shanties which were nothing more than popular songs fitted with new words. Folk-song enthusiasts may object to this, but it is obvious that no collection of shanties could be considered representative which excluded tunes so *universally* employed at sea as 'John Brown's body', 'The Banks of Sacramento', 'Can't you dance the polka?', etc.

Woodstock, July, 1926. R. R. TERRY

NOTES ON THE SHANTIES

1. THE BLACK BALL LINE

This is one of the best known of the older shanties, and some form of it appears in nearly every collection. Strangely enough, Capt. Whall does not include it in his *Sea Songs, Ships, and Shanties*. The version which appears here was sung to me by Mr. Geo. Vickers, in 1914.

2. ONE MORE DAY

This was a homeward-bound shanty in which all the grievances of the voyage were ventilated. It was known to every sailor. The lugubrious manner of its rendering seemed in strange contrast with the elation one expected on nearing port. Whall gives a version on page 77.

3. A-ROVING. I.

I learnt this version from Mr. Jas. Runciman. It differs but little from the usual one found in *The Scottish Students' Song Book* and similar publications. Whall gives a version on page 81.

4. A-ROVING. II.

This version was sung to me by Mr. Short at Watchet, Somerset. There is another version in print (which differs at several points) taken down from his singing. This only goes to prove (what every collector of shanties knows) that shantymen are given to varying their versions according to the mood of the moment.

5. THE BANKS OF SACRAMENTO

This will be recognized as a variant of Stephen Foster's American 'nigger' song, 'Camptown Races' (better known as 'Doodah-doodah-day'), but whether Foster got his tune from the shanty or *vice versa* must remain a moot point. Miss C. Fox-Smith (the well-known nautical authoress) says: 'As a matter of fact, it is a question which of the two (i.e. the song or the shanty) is the older. "The Banks of Sacramento" certainly dates from the late 'forties or early 'fifties; whether "Camptown Races" came earlier than that I cannot say, but I should doubt it.' A biography of Foster gives 1850 as the date of his song, and although this does not preclude the possibility of the shanty being older than the song, neither does it establish it. It is worth asking, however, that if the sailor (always ready as he was to adapt any shore song as a shanty) did seize on 'Camptown Races' because of its amazing popularity, why did he not seize on any other songs of Foster ('Swanee River' for instance) which had a greater vogue and were equally adaptable as shanties?

Whall gives a version on page 65.

6. THE SHAVER

I learnt this shanty from the singing of the late Mr. James Runciman, who told me that he learnt it from a relative who was a great-uncle of mine. It has the same tune as the shanty, 'Poor Paddy works on the railway.' Bullen and others have rejected 'Poor Paddy' on the ground that it was a Christy Minstrel song, and not a real shanty. This is doubtless true as regards the words. But—according to my great-uncle—the tune was sung at sea (to the words of 'The Shaver') before the Christy Minstrels came into existence. Only the first two verses of this shanty are possible in their original form.

7. PADDY WORKS ON THE RAILWAY

This form of the shanty is the best-known amongst sailors. The shantyman always began with 'eighteen hundred and forty-one' and took the following years *seriatim* in successive verses. This shanty differed from most others inasmuch as the couplets always rhymed, and the tune varied less (in the hands of different shantymen) than any other I know. The present words are all from Mr. Short's singing. They are very much like those given by Whall (page 88).

8. CAN'T YOU DANCE THE POLKA?

This was a prime favourite in the palmy days of the sailing vessel. Every sailor knew it. Whall gives a version on Page 65, from which the present solo verses are taken, but the tune (which differs at many points from Capt. Whall's version) is the version sung to me in 1914 by Capt. Robertson.

In childhood I have heard the last lines of the chorus as:

> 'Oh, you New York gels,
> I love you for your money.'

Since 'money' rhymes with 'honey', and 'polka' does not, I am still wondering which was the original.

9. JOHN BROWN'S BODY

The history of this sông has often been told and needs no repetition. For the British sailor it had none of the associations that it held for Americans. The British sailor liked the tune and 'made it into a shanty' which became one of the most popular in his repertory. The words were distinctly ribald, but one must remember that to him 'John Brown' was no more than a figure of speech, as abstract as 'Reuben Ranzo.' The shantyman's historical irrelevancy is seen in verse 2, where 'John Brown' is substituted for 'Jeff Davis' of the original, and the sentence is put into the past tense. The present version was sung to me by Capt. Robertson, but I have rarely met a sailor who did not know some form of it. When the sailor took a shore melody he never debased it; his alterations were usually improvements, and I think this is a case in point.

10. WHOOP JAMBOREE

I have never heard this shanty from anyone save Mr. Short. A version of it, taken down from his singing, has already appeared in print under the title 'Whip Jamboree.' The word as 'coughed up' by Mr. Short (with a shock of the glottis) sounded more like 'Whup.' The printed version gives

but I have set down the actual notes sung by Mr. Short to me. The same version gives three verses; I have set down the *four* which Mr. Short sang.

11. MY JOHNNY

I never heard this shanty save from Mr. Jas. Runciman, and should have set it down as a shore song but for his telling me that he had heard it as a shanty. The words are of the sentimental type beloved of sailors, and the tune is redolent of the Venetian gondola, and the tinkling guitar with its persistent tonic and dominant harmonies (which last I have done my best to avoid).

12. THE DRUMMER AND THE COOK

This is obviously a music-hall song taken over wholesale. I learnt it from Capt. John Runciman, who in turn had it from the cook of the Blyth brig 'Northumberland', in which vessel it was used as a shanty. I remember nothing of this cook except that he was called 'Alf', and that (as was sometimes the case in ships with small crews) he acted as shantyman in the 'Northumberland.' As Capt. John Runciman (who used to sing the whole song) is dead, and as neither Sir Walter Runciman (who also knew it in his youth) nor myself can now remember more than the first verse, I have been guilty of writing the remaining ones which here appear.

N.B. There were two Blyth brigs called *Northumberland*. The first (172 tons) was built at Perth in 1859; the second (271 tons) was built at Blyth in 1862. Information concerning both vessels occurs in Sir Walter Runciman's book *Collier Brigs and their Sailors*. It was in the larger of the two that 'Alf' sailed.

13. MISS LUCY LONG

This was sung to me by Mr. Short. I have never heard it from anyone else.

14. DO LET ME GO, GIRLS

This also was sung to me by Mr. Short. As he had only one verse of words, I have perpetrated the remaining two.

15. BLOW, YE WINDS OF MORNING

This shanty is peculiar. It is the only example I have met of a *sea song* being used as a shanty. Shore songs were annexed wholesale, but the sailor was rigid in banning sea songs for shantying purposes. But Mr. Short, who sang this version to me, assured me that it was used as a shanty in his ship. To my mind, his tune is an improvement on the original, for which see Whall, page 35. Capt. Whall adds the information that it was a song of the midshipman's berth rather than the forecastle, and as he served as midshipman in the Blackwall frigates, he speaks with authority on the point. Mr. Short's words resemble the originals only in the first verse.

16. FIRE DOWN BELOW

This was a shanty known to almost every sailor who had ever worked at the pumps. It had endless verses, most of which I have long forgotten. The authoress of *Roll and Go* describes it well: 'Pumping ship was a long, monotonous spell of hard work unless enlivened by a song. Almost any of the capstan shanties could be used on the pump-brakes, but a few (this one among them) were kept by the force of convention for no other use. Jack would have his joke, even about that most dreaded of dangers—fire at sea; and the joke lay in his choosing non-inflammable portions of the ship in which to locate his imaginary fire. There is always, of course, a fire in the galley, which is the ship's kitchen.'

17. SHALLOW BROWN

This beautiful shanty was a general favourite, and the present version is the one sung by all my sailor relatives. It differs very slightly from that given by Whall, page 119.

18. A LONG TIME AGO

This was another well-known shanty. The version is that sung to me by Mr. Geo. Vickers. As verse 2 never seemed to be omitted, it points to the shanty being of American origin. Masefield quotes a version of the words in *A Sailor's Garland*, but in place of Mr. Vicker's 'family' who 'lived on a hill,' he puts the Yankee Packet:

She was waiting for a fair wind to get under way,
If she hasn't had a fair wind she's lying there still.

In fact, at this point, every version had a verse concerning some person (or persons) or thing that was *static* if certain named conditions were not fulfilled. One, sent me by an old sailor began:

'There once was a farmer in Norfolk did dwell'

As his rhyme for 'dwell' happened to be 'hell', it is easy to see what alternative dwellings were open to the farmer.

19. WON'T YOU GO MY WAY

This charming shanty was sung to me by Mr. Short. I have not met any other sailor who knows it. A version (differing from the present one in the music of bar 9, and the words of verses five and six) is given in C. J. Sharp's collection, taken down from Mr. Short's singing, also. Mr. Short may have exercised the shantyman's privilege of varying melody or words at will. At any rate, I have set both down as he sang them *to me*.

20. HILO JOHN BROWN

Whall gives a version of this (p. 85) under the title, 'Stand to your ground', in which the words differ only slightly from those I have heard. As my version of the tune is an inferior one I am enabled, by the kind permission of Capt. Whall's executors, to reproduce his melody with one difference: Capt. Whall gives a G sharp in bars 2 and 3 of the last line, and this is, no doubt, the way it was sung at sea in his time. But the tune is not in a minor *key* but in the First *Mode*. I have, therefore, eliminated the G sharp. I feel justified in this course because—although I have met only two seamen who knew the shanty, both sang the chorus with the G natural.

21. ROLL THE COTTON DOWN

This was known to every seaman who had been in the cotton trade. All my sailor relatives sang some form of it. The present version is that of Capt. John Runciman. It is clearly of American origin.

22. ROUND THE CORNER SALLY

I have not heard anyone save Mr. Short sing this shanty. The first verse, as I took it down from him, had three lines for the soloist. As I knew of only one other hauling shanty with this peculiarity ('Cheer'ly men') I bided my time until Mr. Short had sung other verses. I then found that these verses were in *couplets* (the usual hauling form). I have, therefore, adhered to the couplet form throughout.

23. THE BULLY BOAT IS COMING

This was sung to me by Mr. Short. His words run, 'Don't you hear the paddles rolling', but Mr. Mayoss, Mr. Allen, and others always sang 'paddles roaring', which seems the more probable reading.

I suspect 'Rando' ought to have been 'Ranzo', but as Mr. Short sang the former word, I have set it down here.

24. MY TOMMY'S GONE AWAY

This is a variant of the sentiment of 'Tom's gone to Hilo' (see Part I) but the tune is different and not so good. The version is that of Mr. Short.

25. SING FARE YOU WELL

This was also sung to me by Mr. Short. I had not heard it before, nor does it appear in any other collection.

26. O BILLY RILEY

Sung to me by Mr. Short. I have not found any other sailor who knows it.

27. TIME FOR US TO LEAVE HER

The original words (for which see Whall, p. 68) date from the 'fifties, and referred to the Irish emigrants whom the old packet-ships carried to America. They soon fell into disuse, and the shanty was used for ventilating grievances when nearing a home port. The tune was known to every sailor, and appears in most collections in one form or another. Its use seems to have been confined to British ships. I have as yet found no trace of its being used in American vessels. For further information see Whall, page 68.

28. LIZER LEE

Sung to me by Mr. Short. It is a better version than those sung by Sir Walter Runciman and others. The best version I know is the printed one by Frank Bullen, but as I never heard Mr. Bullen sing it (and in this book I have set down no tune which has not actually been sung to me by some sailor) I have done 'the next best thing' and given Mr. Short's version. Bullen's capital tune runs thus:

29. A HUNDRED YEARS ON THE EASTERN SHORE

Some form of this was known to nearly every British seaman. The present version was sung to me by Mr. Geo. Vickers. Joanna C. Colcord (in her book, *Roll and Go*) states (I do not know on what authority) that it is 'the only shanty which can be identified with the Baltimore clippers.' She gives four additional stanzas, which I do not include here, as I have not heard any British seaman sing them. They were sung in American vessels, however.

30. WALK HIM ALONG, JOHNNY

I have heard no one sing this save Mr. Short. The tune differs at several points (notably, bars 6 and 7, page 59) from C. J. Sharp's printed version taken down from Mr. Short. But I have set it down exactly as he sang it to me.

31. HILONDAY

I learnt this in boyhood from the late Mr. James Runciman. I do not know in which ship he picked it up, but one of my earliest recollections is hearing him and W. E. Henley give tongue to it at the house of the latter (in the days when he lived at Shepherd's Bush—then an outlying suburb). Henley's knowledge of the sea (like R. L. Stevenson's) was the acquired knowledge of the literary landsman, but shanties—especially the grim ones—had a special appeal for him, and he was fond of singing them. The sea song, 'Time for us to go', which he incorporated in the play of *Admiral Guinea* (calling it a 'chanty') I learnt from him in my boyhood, to a tune which I understood was his own composition. It is a good imitation of a capstan shanty, but I do not include it in this collection as it was never sung at sea; I hope one day to publish it separately.

32. STORMALONG

This (like No. 10 in Part I) is one of the many shanties which mourn the mythical hero 'Stormy.' It is the one which was the most popular, and every seaman knew it. The present version is that of Sir Walter Runciman. Whall gives a version on page 87.

33. SO HANDY, MY GELS

Sung to me by Mr. Short, Mr. Morley Roberts, and most other sailors of my acquaintance. See also Whall, page 128.

34. THE SAILOR LIKES HIS BOTTLE, O

Although I have known this shanty almost as long as I can remember, I have never heard it aboard ship, and so I do not pretend to explain how the first line was manipulated. At the moment of writing I cannot get in touch with the particular seamen whom I know could enlighten me. Rather than delay the publication of this collection I will hold the matter over, and publish the result of my inquiries in my forthcoming book (referred to in the introduction) on the shanty.

35. HAUL AWAY, JOE

This major version of the fine minor tune in Part I (page 56) was almost equally popular. I noted that Sir Walter Runciman, and most of the older generation of seamen always sang the minor version. The major one was mainly confined to a younger generation, but Whall gives it on page 117.

THE SHANTY BOOK.

PART II.

1. The Black Ball Line.

(CAPSTAN)

2. It'll carry you along through frost and snow,
And take you where the wind don't blow.

3. At Liverpool Docks I bade adieu
To Poll and Bet, and lovely Sue.

4. And now we're bound for New York Town,
It's there we'll drink, and sorrow drown.

5. It's there I'll sport my long-tailed blue. *(twice)*

2.

One more day.

(CAPSTAN.)

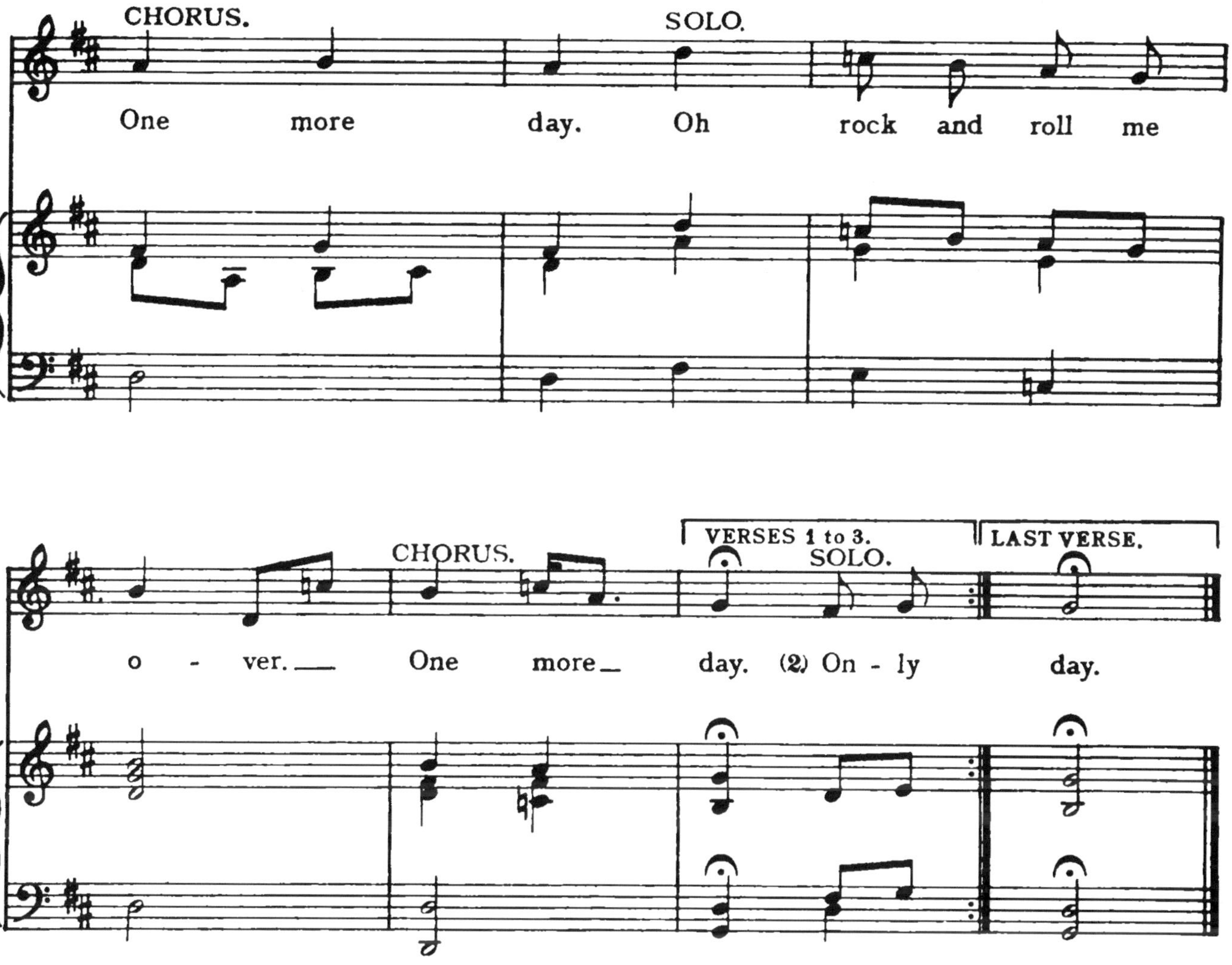

2. Only one more day, my Johnny;
One more day.
We'll cross the bar to-morrow,
One more day.
(Repeat whole of first verse as chorus.)

3. Don't you hear the old man roarin', Johnny,
One more day?
Don't you hear that pilot bawlin',
One more day?
(Repeat first verse as before.)

4. Can't you hear those gals a-callin', Johnny,
One more day?
Can't you hear the capstan pawlin',
One more day?
(Repeat first verse as before)

3.

A-Roving. I.

(CAPSTAN.)

2. I took that fair maid for a walk.
Mark well, *etc.*
I took that fair maid for a walk.
And we had such a loving talk.
I'll go no more, *etc.*

3. I put my arm around her waist.
Mark well *etc.*
I put my arm around her waist,
So slim, and trim, and tightly laced.
I'll go no more *etc.*

4. I took that maid upon my knee.
Mark well, *etc.*
I took that maid upon my knee.
Said she, "Give over! Let me be!"
I'll go no more, *etc.*

4. A-roving. II.

(CAPSTAN.)

2. I took this fair maid for a walk,
   Bless you etc.
   And we had such a loving talk.
   I'll go no more etc.

3. I took her hand within my own,
   And said "I'm bound for my old home."

4. In Plymouth Town there lived a maid,
   And she was mistress of the trade.

5. The banks of Sacramento.

(CAPSTAN.)

2. We're bound away at the break of day. *(twice)*

3. The rose is red; the violet's blue;
O Amble girls we all love you.

4. Sally Brown she's come to town.
Sally Brown's got a new silk gown.

6.

The Shaver.

(CAPSTAN.)

2. O they whacked me up, and they whacked me down;
The mate he cracked me on the crown;
They whacked me round and round and round,
When I was, *etc.*

3. When I went aloft by the lubber's hole,
The mate he cried "O dang yer soul,
It's the futtock shrouds is the way yer bound,"
When I was, *etc.*

4. When we lollop'd around about Cape Horn,
I wished that I had never been born,
And I wished I was home all safe and sound,
When I was, *etc.*

7. Paddy works on the railway.*

(CAPSTAN.)

* A "Christy Minstrel" version of the music of the preceding. See notes on the Shanties.

2. In eighteen hundred and forty two
I did not know what I should do.
And I resolved to put her through
To work, *etc.*

3. In eighteen hundred and forty three
I paid my passage across the sea,
To New York, and Amerikee
To work, *etc.*

4. In eighteen hundred and forty four
I landed on the American shore,
And never to return no more
To work, *etc.*

5. In eighteen hundred and forty five
Things looked pretty well alive,
And I thought to myself I'd strive
To work, *etc.*

6. In eighteen hundred and forty six,
When I was in a terrible fix,
I thought to myself I'd take my sticks,
To work, *etc.*

7. I had a sister, her name was Grace,
Bad cess unto her ugly face,
She brought me to a deep disgrace
A-working, *etc.*

8. Can't you dance the Polka.

(CAPSTAN.)

2. To Tiffany's I took her,
  I did not mind expense;
I bought her two gold earrings,
  And they cost me fifty cents.
      Then away, *etc.*

3. Says she "You lime-juice sailor,
  Now see me home you may."
But when we reached her cottage door
  She unto me did say—
      Then away, *etc.*

4. My flash man he is a Yankee,
  With his hair cut short behind;
He wears a tarry jumper,
  And he sails in the Black Ball Line.
      Then away, *etc.*

9.

John Brown's body.

(CAPSTAN.)

2. We hanged John Brown upon a sour apple tree *(thrice)*

3. John Brown's wife has got a wart upon her nose *(thrice)*

4. John Brown's daughter chews terbacker by the pound *(thrice)*

5. John Brown's baby is a yankee-doodle-doo *(thrice)*

6. John Brown's body lays a-mould'ring in the grave *(thrice)*

10. Whoop Jamboree.

(CAPSTAN.)

2. Now Cape Clear it is in sight,
We'll be off Holyhead by to-morrow night;
And we'll shape our course for the Rock Light,
O Jenny, *etc.*

3. Now, my lads, we're round the Rock,
All hammocks lashed and chests all locked.
We'll haul her into the Waterloo dock.
O Jenny, *etc.*

4. Now, my lads, we're all in dock,
We'll be off to Dan Lowrie's on the spot;
And now we'll have a good roundabout.
O Jenny, *etc.*

11.

My Johnny.

(CAPSTAN.)

2. So gay we went away,
Me and my pretty Johnny,
My own dear Johnny, *etc.*
But where is he to-day?
O always think of Johnny.
My lively Johnny, good-bye.
'Twas just by Finisterre
Where the birds are free in the air,
We buried Johnny there,
We buried Johnny there,
In the middle of the sea, *etc.*

12. The drummer and the cook.

(CAPSTAN.)

2. When this couple went a-courtin' for to walk along the shore,
Sez the drummer to the cookie, "You're the gel that I adore."

3. When this couple went a-courtin,' for to walk along the pier,
Sez the cookie to the drummer "An' I love you too, my dear."

4. Sez the drummer to the cookie, "Aint the weather fine to-day?"
Sez the cookie to the drummer, "Is that all ye got to say?"

5. Sez the drummer to the cookie, "Will I buy the weddin' ring?"
Sez the cookie "Now you're talkin'. That would be the very thing."

6. Sez the drummer to the cookie, "Will ye name the weddin' day?"
Sez the cookie, "We'll be married in the merry month o' May."

7. When they went to church to say "I will," the drummer got a nark*
For her one eye gliffed✝ the Parson, and the t'other killed the Clerk.

*Nark = a disagreeable surprise caused by a *person*, not by a circumstance.

✝To gliff = to frighten.

13.

Miss Lucy Long.

(CAPSTAN.)

2. O! as I walked out one morning fair,
To view the views and take the air
To me way *etc.*

3. 'Twas there I met Miss Lucy fair;
'Twas there we met I do declare.
To me way *etc.*

4. I rung her all night and I rung her all day,
And I rung her before she went away.
To me way *etc.*

14. Do let me go, girls.

(CAPSTAN.)

2. A-courting of the maiden came a sailor long ago.

3. But he left her broken-hearted on the shores of Callao.

15. *Blow, ye winds of morning.

(CAPSTAN.)

*This is the only instance– in my experience– of a Sea Song being adapted and used as a Shanty. [Ed.]

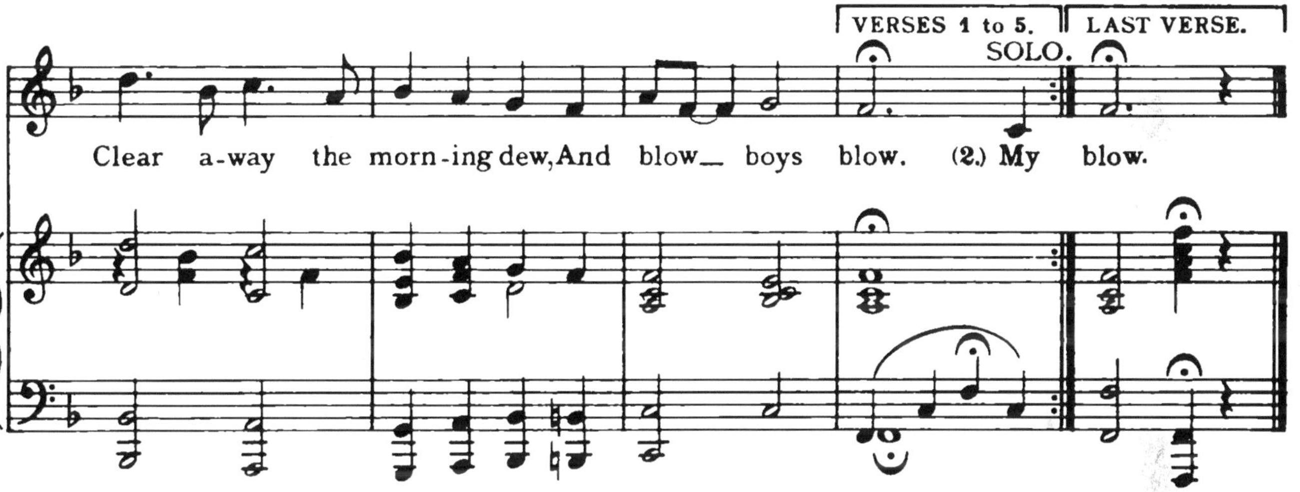

2. My father has a milk-white steed,
And he is in the stall.
He will not eat his hay or corn,
Nor will not go at all.

3. When we goes in a farmer's yard
And sees a flock of geese,
We dang their eyes and cuss their thighs
And knock down five or six.

4. As I was a walking
Down by the riverside
It's there I saw a lady fair
A-bathing in the tide.

5. As I was a-walking
Out by the moonlight,
It's there I saw a yeller gel,
And her eyes they shone so bright.

6. As I was a-walking
Down Paradise Street,
It's there I met old John de Goss,*
He said "Will you stand treat?"

* The reference is not to the famous baritone singer, but to a Liverpool shipowner of last century (with the shellback's mispronunciation of his name, of course).

16.

Fire down below.

(PUMPING-SHIP SHANTY.)

CHORUS.

2. Fire in the fore-top, fire in the main;
It's fetch a bucket o' water girls, and put it out again.

3. Fire in the fore-peak, fire down below;
It's fetch a bucket o' water girls, there's fire down below.

4. Fire in the windlass, fire in the chain;
It's fetch a bucket o' water girls, and put it out again.

5. Fire up aloft, and fire down below;
It's fetch a bucket o' water girls, there's fire down below.

17. Shallow Brown.

(HALLIARDS.)

2. Away, I'm bound to leave you.
I never will deceive you.

3. I love to look upon you.
I bet my money on you.

4. O, Shallow in the morning,
Just as the day was dawning.

5. Oh put my clothes in order;
I'm bound across the border.

18.

A long time ago.

(HALLIARDS.)

2. Away down South where I was born,
   Among the fields of golden corn.

3. A Yankee packet lay out in the bay,
   A-waiting a fair wind to get under weigh.

4. There once was a family lived on a hill,
   And if they're not dead they're all living there still.

19.

Won't you go my way?

(HALLIARDS.)

2. In the morning bright and early.

3. O Juliar, Ann, Mariar.

4. I asked that girl to marry.

5. O marry, do not tarry.

6. She said she'd rather tarry.

20. Hilo, John Brown.

(HALLIARDS.)

2. Sally she a 'Badian bright Mulatter;
Sally pretty gal, but can't get at her.

3. Seven long years I courted Sally;
Sally she would flirt but nebber marry.

4. Stand to your ground and walk him up lively,
Or de mate come around a-dingin' and a-dangin'.

21. Roll the cotton down.

(HALLIARDS.)

2. In the morning bright and early *(twice)*

3. So early in the morning
Before the day was dawning.

4. We'll screw him up so cheerly *(twice)*

5. I'm off across the border *(twice)*

6. Farewell, I'm bound to leave you;
I never will deceive you.

22. Round the corner Sally.

(HALLIARDS.)

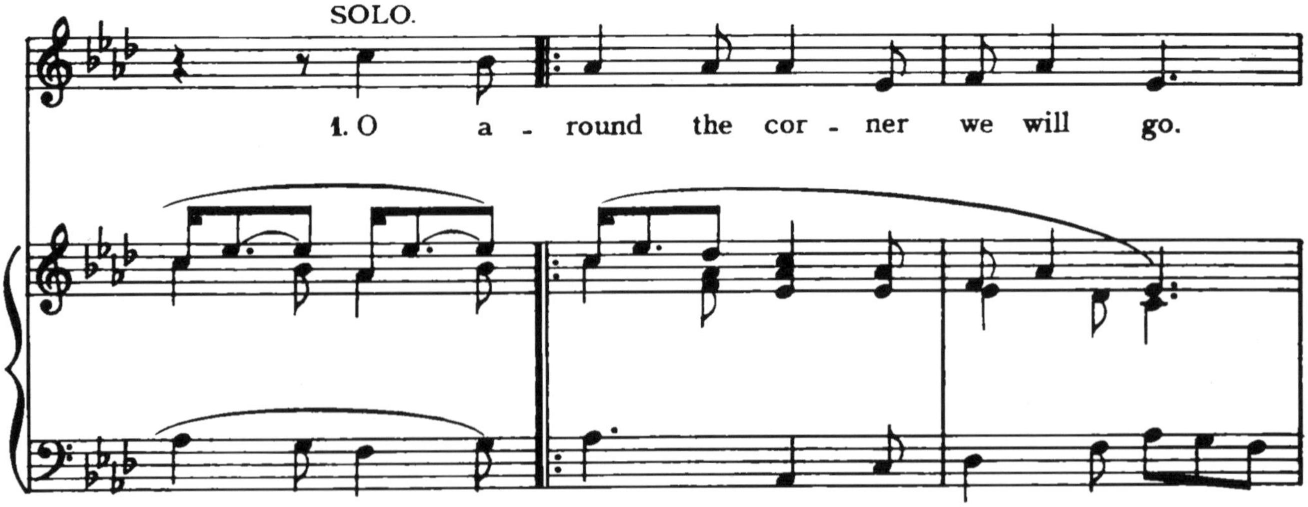

2. To Madam Gashee's we all will go,
For Madamoiselle you all do know.

3. O Madamoiselle we'll take her in tow;
We'll take her in tow to Callao

4. O I wish I was at Madam Gashee's;
It's there we'll sit and take our ease.

23. The bully boat is coming.

(HALLIARDS.)

2. O the bully boat is coming,
Down the Mississippi floating. } *(twice)*

3. Oh I'm bound away to leave you,
And I never will deceive you. } *(twice)*

4. When I come again to meet you,
It's with kisses I will greet you. } *(twice)*

5. Oh the bully boat is coming,
Don't you hear the paddles rolling? } *(twice)*

24. My Tommy's gone away.

(HALLIARDS.)

2. Tommy's gone to Liverpool,
To Liverpool, that noted school.

3. Tommy's gone to Baltimore,
Oh Tommy's gone to Baltimore.

4. Tommy's gone to Mobile Bay,
To screw the cotton by the day.

5. Tommy's gone for evermore,
Oh Tommy's gone for evermore.

25. Sing fare you well.

(HALLIARDS.)

2. O fare you well my bonny young gel *(twice)*

3. As I walked out one morning fair
It's there I met a lady fair.

4. At her I winked I do declare *(twice)*

5. Up aloft this yard must go *(twice)*

6. I thought I heard the skipper say,
One more pull and then belay.

7. Fare you well, I wish you well;
Fare you well till I return.

***This note will be used only in Verses 3 & 4.**

26. O Billy Riley.

(HALLIARDS.)

2. O Mister Riley, Ö Missus Riley *(twice)*

3. O Missy Riley, little Missy Riley *(twice)*

4. O Missy Riley, screw him up so cheer'ly *(twice)*

27. Time for us to leave her.

(HALLIARDS.)

2. Me'og'ny beef and weevill'd bread!
   I wish old Weather-phiz was dead.

3. The rain it rains the 'ole day long;
   The Nor'-East wind is blowin'-strong.

4. It's pump or drown the old man said. *(twice)*

5. I thought I heard the captain say
   "To-morrow you shall have your pay."

6. O what will us poor shellbacks do? *(twice)*

7. We'll pack up our traps and go on shore. *(twice)*

8. O times is hard and wages low. *(twice)*

28.

Lizer Lee.

(HALLIARDS.)

2. Oh Lizer Lee she slighted me;
Now she will not marry me.

3. When I sailed across the sea,
Lizer said she'd be true to me.

4. I promised her a golden ring. *(twice)*

5. Up aloft this yard must go,
Mister Mate he told us so.

6. I thought I heard the skipper say,
"One more pull and then belay."

29. A hundred years on the eastern shore.

(HALLIARDS.)

2. In the Black Ball Line I served my time. *(twice)*

3. A hundred years is a very long time. *(twice)*

4. A hundred years have passed and gone. *(twice)*

5. A hundred years will come once more. *(twice)*

30. Walk him along, Johnny.

(HALLIARDS.)

SOLO.
bur-y-ing ground. Then a-way - ay you Storm - y,
CHORUS.
Walk him a-long, John-ny, car-ry him a-long.
SOLO.
Way - ay you
Storm - y,
CHORUS.
Car-ry him to the bur-y-ing ground.

31.

Hilonday.

(HALLIARDS.)

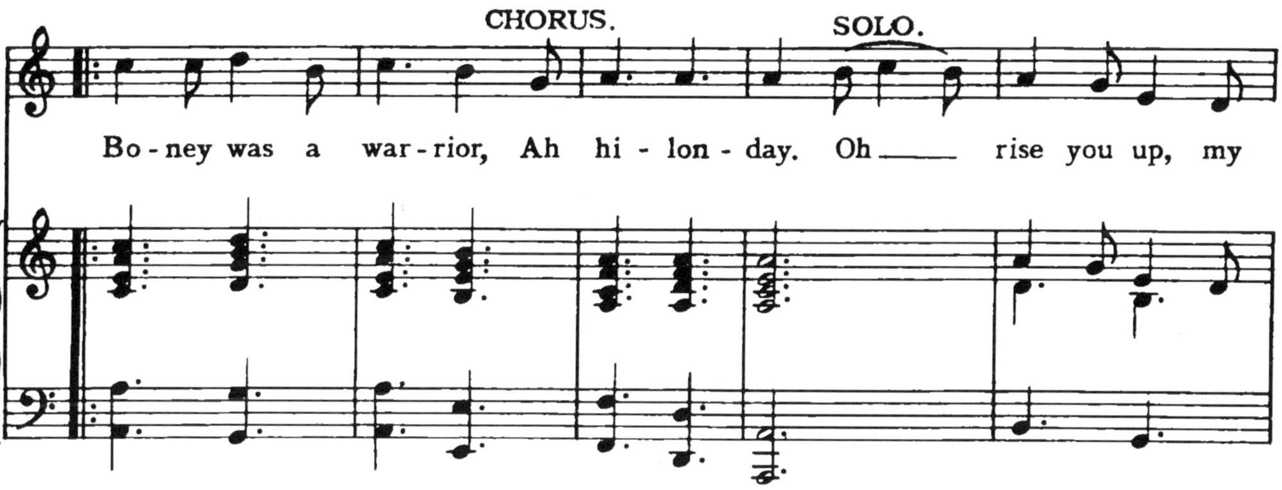

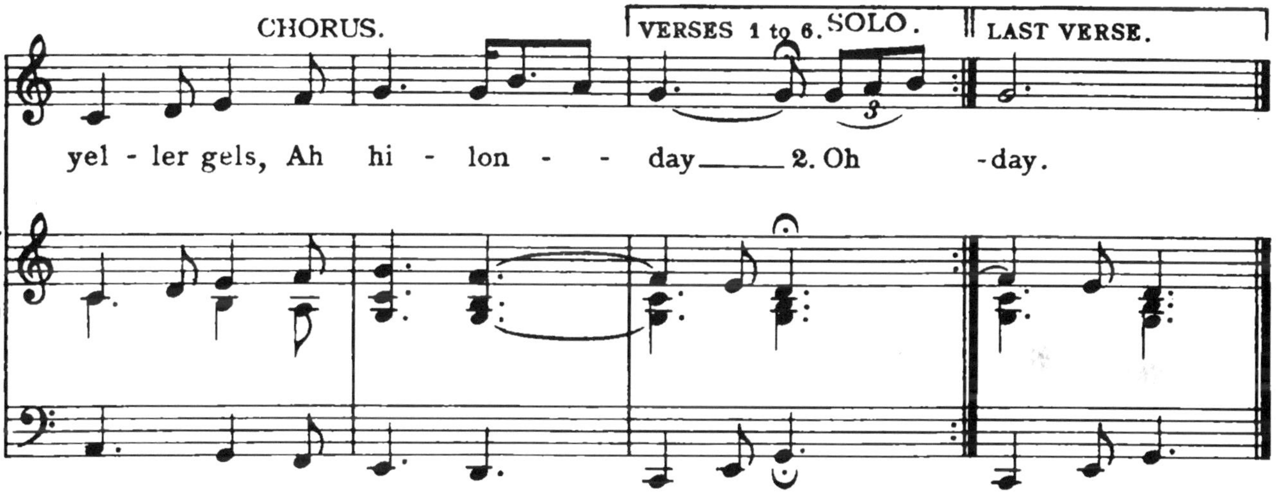

2. Oh Boney beat the Rooshans.
   Oh Boney beat the Prooshans.

3. Oh Boney went to Moscow. *(twice)*

4. Oh Moscow was afire. *(twice)*

5. Oh Boney was defeated. *(twice)*

6. Oh drive her, captain, drive her. *(twice)*

7. Oh captain, make her nose blood. *(twice)*

32. Stormalong.

(HALLIARDS.)

2. I dug his grave with a silver spade.

3. I lowered him down with a golden chain.

4. We carried him away tö Mobile Bay.

5. We'll never see hi̇s like again.

6. Stormy was ä good old man.

7. Stormy he i̇s dead and gone.

33. So handy me gels.

(HALLIARDS.)

2. Be handy with your washing, girls,
Because my love's a dandy, O.

3. My love she is a dandy, O,
And she is fond of brandy, O.

4. O shake her up and away we'll go;
Up aloft from down below.

34. The Sailor likes his bottle, O.

(INTERCHANGEABLE SHANTY.)

2. So early in the morning,
The sailor likes his baccy, O.
A packet o' shag, and a packet o' twist,
And a packet o' Yankee Doodle, O.
Chorus.

3. So early in the morning,
The sailor likes the lasses, O.
The lasses o' Blyth, and the lasses o' Shields,
And the lasses across the water, O.
Chorus.

35.

Haul away, Joe. II.

(FORE-SHEET OR SWEATING-UP.)

2. Oh once I had a nigger gel, and she was fat and läzy.

3. And then I had an Irish gel; she nearly druv me cräzy.

4. King Louis was the king o' France before the revolütion.

5. King Louis got his head cut off and spoiled his constitütion.

N. B. Any of the verses of "Haul away, Joe" No. I. ("The Shanty Book Part I" page 56) can be sung to this Shanty.

Made in United States
Troutdale, OR
11/03/2026

56845093R00046